# Topical Conference: Sensors for Sustainability



Held at the 2023 AIChE Annual Meeting

Orlando, Florida, USA
5-10 November 2023

ISBN: 978-1-7138-9315-8

**Printed from e-media with permission by:**

Curran Associates, Inc.
57 Morehouse Lane
Red Hook, NY 12571



AIChE
120 Wall Street, FL 23
New York, NY 10005-4020

Phone: (800) 242-4363
Fax: (203) 775-5177

www.aiche.org

**Additional copies of this publication are available from:**

Curran Associates, Inc.
57 Morehouse Lane
Red Hook, NY 12571 USA
Phone: 845-758-0400
Fax: 845-758-2633
Email: curran@proceedings.com
Web: www.proceedings.com

# TABLE OF CONTENTS

## FIELD-DEPLOYABLE SENSORS



## GENERAL POSTER SESSION IN SENSORS

## NEXT-GEN SENSORS



## SENSORS AND MONITORING FOR HEALTH

**STUDENT COMPETITION IN SENSORS (SPONSORED)**

# 213a Novel Field Deployable Sensors to Monitor How Dynamic Hydrology Shapes Nutrient and Element Transformations in a Great Lakes Coastal Estuary: A Two-Year Study

*Sai Prasanna Chinthala, Bukola Adesanmi, Zachary Cheney, Erin Hasset, Erin Eberhard, Lauren Kinsman-Costello, John Senko, Timothy Morin, Thomas Quick, Chelsea Monty-Bromer*

The propensity for redox processes to occur is often predicted using measured or assumed redox potential ($E_h$). However, process measurements often deviate from predictions based on the traditional, thermodynamic "redox tower" paradigm. Previous work has shown that poorly measured, granular scale spatial and temporal soil heterogeneity causes apparent departures from thermodynamic exclusion principles at bulk scales, with consequences for large-scale biogeochemical cycling which are not currently resolved in ecosystem, regional, or global scale models.

To help study these processes at a smaller scale, we developed a 3D printed graphite electrode array to perform zero resistance ammetry (ZRA) measurements to measure redox disequilibria at various spatial scales in benthic sediments. This allowed us to detect the distributions, extents, and kinetics of biogeochemical processes occurring in near real-time. ZRA was used to measure electrical current that arose from microbially-induced redox disequilibrium. Prior to deployment in the field, we characterized the 3D printed electrodes using cyclic voltammetry and chronocoulometry. We also used the 3D printed graphite electrodes in laboratory-scale experiments mimicking various posions on the redox ladder using a split cell setup containing poorly crystalline Fe(III) (hydr)oxide-rich soils. When both chambers were incubated under oxic or anoxic conditions, no current was observed. However, when one chamber was oxic and the other anoxic, no Fe(III) reduction was observed in the oxic chamber and Fe(II) accumulated in the anoxic chamber. The redox disequilibrium generated between chambers was accompanied by electron transfer from the anoxic to the oxic chamber detected as ZRA.

We then refined the design, construction, and deployment of a paired ZRA and Eh multi-sensor system to detect electrochemical signals across and directly beneath the sediment-water interface at nested scales (sub-millimeter to decimeter) at a shallow location (< 10 cm surface water) in the Old Woman Creek wetland. During the summers of 2022 and 2023, we deployed these sensors and collected concurrent data on dissolved oxygen dynamics, surface and pore water nutrient concentrations, greenhouse gas fluxes (chamber measurements), and soil geochemistry. Future work will use this data to demonstrate the feasibility of integrating microsite electrochemical and redox variability to the ecosys ecosystem scale model for improved representation of soil redox processes in spatially variable and temporally fluctuating systems. Field data shows the success of a vertically stacked ZRA electrode at monitoring microbial processes at the small-scale.

## 213b Engineered, Field-Deployable Biosensors for Detection of Pfas Compounds in Soil, Water and Food

*Madison Mann, Victoria Kartseva, Chelli Stanley, Bryan Berger*

Per- and polyfluoroalkyl substances (PFAS) are a large group of synthetic fluorinated chemicals with surface active and water-repellent properties. Given their wide-spread use in numerous consumer and industrial products and persistence in the environment due to strong carbon-fluorine bonds, there is significant accumulation of PFAS in water, soils, food and ultimately people. Recent studies have indicated that any level of PFAS exposure is detrimental to human health; currently, the EPA has set legal exposure limits for PFOA and PFOS at 4 ppt, although higher levels are often found in water, soil and food. As most human interaction with PFAS comes from ingestion, it is important to be able to detect PFAS in multiple types of heterogeneous samples. We present an approach to designing fluorescence-based biosensors for the rapid detection of PFAS based on human liver fatty acid binding protein (hLFABP). In our first approach, introduction of solvatochromic fluorophores within the ligand binding pocket (F50) allowed for intrinsic detection of PFOA, PFOS, and PFHxS via blue-shifts in fluorescence emission spectra. Initially, a single tryptophan mutation (F50W) was found to be able to detect PFOA with a LOD of 2.8 ppm. We improved the sensitivity of the biosensor by exchanging tryptophan for the thiol reactive fluorophore, acrylodan. The acrylodan conjugated C69S/F50C hLFABP variant is capable of detecting PFOA, PFOS, and PFHxS in PBS with LODs of 112 ppb, 345 ppb, and 1.09 ppm respectively. The protein-based sensor is also capable of detecting these contaminants at similar ranges in spiked environmental water samples, including samples containing interfering anionic surfactants such as sodium dodecyl sulfate (SDS). In our second approach, we developed a circularly-permuted green fluorescent protein (GFP)-based biosensor using hLFABP as the binding element to implement a genetically-encoded biosensor for cell-based detection; initial designs indicate our GFP biosensor can also detect key PFAS compounds such as PFOA. Additional work demonstrates these biosensors can be expanded to detect other co-pollutants such as fuel hydrocarbons. Ongoing work is focused on optimizing signal-to-noise ratio and lowering LOD for PFAS detection, as well as expanding testing to other foods. Overall, this work demonstrates engineered hLFABP is a useful platform for detection of PFAS in environmental water samples and highlights its ease of use and versatility in field applications.

## 213c Development of Measurement Techniques for Multiphase Flows Using Electrical Resistance Tomography

*Weiguo Xie*

Multiphase flows can be seen in many natural phenomena, and also are crucial within various industries. However, it has always been an enormous scientific challenge to quantify properties for multiphase flows due to their complex features and the lack of diagnostic tools. Electrical resistance tomography (ERT) has been applied in a wide range of scientific and engineering fields, such as geophysical study, clinical diagnosis, chemical or mineral process measurement, and process optimization. Typically, ERT is used to obtain information about processes occurring in vessels and pipelines by mounting an array of electrodes, equally distanced, to the outside of a vessel, as shown in Figure 1. There is also another way to place ERT sensor inside a vessel and move it to any location for measurement as shown in Figure 2. Over last two decades, in my research to extract information and data analysis from ERT conductivity distribution images, ERT system has been developed to measure instantaneous local physical quantities for multiphase flows in chemical engineering and mineral processing, including: i) Estimate bubble size in flowing foams; ii) Identify bubble coalescence zones in multiphase flows; iii) Measure turbulence in multiphase flows; iv) Monitor chemical reaction and fluid mixing.

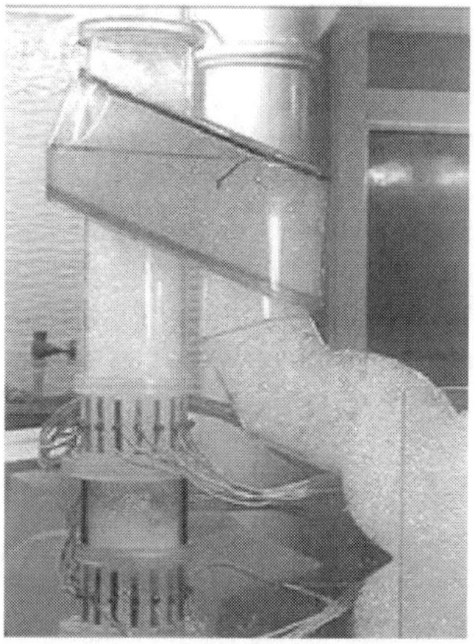

Figure 1: Laboratory flowing foam flotation column with two-layer ERT sensors

Metso 3 m$^3$ RCS flotation test rig    ERT sensor

Figure 2: Turbulence measurements for Metso 3 m$^3$ RCS flotation test rig using ERT

## 213d Development of Continuous Methane Monitors for Detection of Emissions from the Oil and Gas Sector

*Aditi Chakrabarti, A. Ballard Andrews, Mathieu Dauphin, Kashif Rashid, Lukasz Zielinski, Junyi Yuan, Manasi Doshi, Christopher Boucher, Andrew Speck*

Natural gas consists mostly of methane, an invisible and harmful greenhouse gas known to have present global warming impact 50 times greater than that of carbon dioxide. The US Environmental Protection Agency (EPA) has proposed to implement regulatory standards for comprehensive emissions monitoring and mitigation from oil and natural gas facilities. To address the urgency of detecting methane emissions in these facilities, a continuous monitoring solution was developed for quantitative mapping of methane leaks via using point sensors that are installed permanently at the periphery of a site, are IoT-enabled, and run on solar power. Our low-cost sensors use calibration models developed using machine learning that account for environmental effects in the sensor response. The monitoring system collects data in real time on methane emissions and meteorological conditions, which are filtered and averaged via edge computing using atmospheric dispersion statistics. The data are then processed using inversion algorithms to estimate leak rates and their positions. These results are displayed in a digital platform where users can visualize methane leaks mapped on their facilities in real time. Our methane emissions detectors have been deployed at a few different oil and gas facilities, including two test sites where their detection accuracies were optimized by using controlled releases of methane with flow rates from 1 to 15 kg/h for several months. The results show that our point sensors can detect methane concentrations with ppm-level accuracy over a wide range of temperature, humidity, and wind speed and direction. From the concentration data, our interpretation algorithms can estimate leak rates and their positions quantitatively, even for complex field sites. Our methane monitoring system can identify leak rates as low as 0.2 kg/h. Its capability to accurately pinpoint methane leaks with no false positives will enable oil and gas operators to intervene and mitigate their emissions to achieve sustainable and cost-effective operations in low-resource settings.

## 213e Field Deployable Sensor Arrays with Molecular Selectivity for Soil Health

*Yaoli Zhao, Gayatri Ranade, Will Stump, Michael Tidwell, Christopher S Johnson, Chris Swindell, Charles Van Neste, Thomas Thundat*

Monitoring of soil heath is of utmost importance in agriculture. Detecting the concentration of $CO_2$, $CH_4$, $N_2O$, and volatile organic compounds produced during the metabolomic activities of microbiota is an efficient way to monitor soil health. The use of vapor sensors based on microfabricated cantilever arrays is a promising platform for the development of miniature, field-deployable sensors. These array-based sensors are highly sensitive, small in size, capable of multi-analyte detection, and have low power consumption, making them ideal for high-performance sensing. However, challenges such as poor molecular selectivity, especially for trace amounts of small molecules, and lack of reproducibility, have hindered the translation of this sensor platform into practical reality. Conventional methods of achieving selectivity in chemical sensing rely on immobilized receptors or chemical interfaces on sensor surfaces. However, this approach results in poor selectivity for small molecule detection when using reversible receptors. In addition, non-uniformity in the graft density of the immobilized receptor results in unacceptable levels of irreproducibility. To overcome these challenges, we have used a multi-modal, multi-physics approach for generating orthogonal sensor responses. The use of deep learning techniques to analyze multi-modal data shows significant improvements in selectivity, sensitivity, and reliability of these sensors. To deploy these sensors for field-scale studies, we have developed a long-range through-the-soil (TTS) wireless power transfer method. Energy is wirelessly distributed through conduction currents that propagate outward along and through the soil from an adjacent solar panel array. We have demonstrated power transfer within a 0.25-ha (0.6acre) area using an average input power of 6W. The combination of highly sensitive vapor sensors with improved selectivity, multi-modal data analysis, and long-range wireless power transfer offers a promising solution for the efficient and effective monitoring of soil health in agriculture.

## 213f Development of a Colorimetric Sensing Technology for Rapid Volatile Sulfur CompoundClassification

*Oindrila Hossain, Yan Wang, Mingzhuo Li, Sina Jamalzadegan, Noor Mohammad, Aditi Dey Poonam, Qingshan Wei*

Volatile organic compounds (VOCs) are naturally occurring components found in various fruits and vegetables. For vegetables, VOCs are closely associated with their quality attributes, such as firmness, soluble sugars, flavonols, and translucency. For instance, eye irritation and pungent odors are commonly experienced when cutting onions, caused by a number of volatile sulfur compounds and aldehydes present in them. The current standard methods, such as GC-MS, PTR-MS, are inadequate for detecting these volatiles as VOCs are short-lived and invasive sampling is required. To address this issue, we developed a paper-based colorimetric VOC sensor to detect major sulfur VOCs (i.e., sulfoxides, sulfides, mercaptans, thiophenes) without bulky equipment. Our results showed that the colorimetric sensors successfully detected and classified different VOCs, which can be used to distinguish several onion varieties with high classification accuracy. These findings are promising and offer a potential alternative to the current standard method for detecting vegetable VOCs for rapid quality assessment in the field.

# 213g A Nose for Robots: Engineering Atomically Precise Arrays of Olfactory Receptors

*Brian A. Day, Christopher E. Wilmer*

### A Nose for Robots: Engineering Atomically Precise Arrays of Olfactory Receptors

Most gas sensors can only detect a single gas species (e.g., a household carbon monoxide detector), with a small subset able to distinguish among a few gas species (e.g., commercial "electronic nose" sensors). However, few attempts have been made – and none successfully – at creating a device that could detect a very broad spectrum of gases under a wide range of operating conditions. This seemingly challenging task is carried out every day by our own biological noses, however. Creating an artificial nose as capable as a biological nose would have enormous socioeconomic impact; analogous to the invention of the camera (i.e., the "artificial eye"). It would be transformative for a wide range of industrial and military applications, but, in particular, it would revolutionize health diagnostics.

So what would it take to build a better electronic nose? Electronic noses rely on arrays of chemically distinct materials that each interact uniquely with different gas species, which is closely analogous to how biological noses work. Current electronic noses, however, use small arrays with typically less than 20-30 materials. Dogs, on the other hand, have thousands of distinct olfactory receptors, each of which are replicated tens of thousands of times throughout the interior of their noses. Recent evidence suggests that building larger arrays can significantly improve electronic nose performance, but only if the sensing materials are carefully chosen. For large arrays this becomes a classic "big data" problem, requiring significant computational power and clever algorithms to address the combinatorial explosion of possible arrays one could construct.

In this talk, we discuss recent work on the computational design of gas sensing arrays, based on modeling gas adsorption in metal-organic frameworks (MOFs), and lay out a roadmap for how we might eventually build an electronic nose that could rival a dog's one.

## 213h High Sensitivity Saliva-Based Biosensor in Detection of Breast Cancer Biomarkers: HER2 and CA15-3

*Hsiao-Hsuan Wan, Haochen Zhu, Chao-Ching Chiang, Jian-Sian Li, Fan Ren, Cheng-Tse Tsai, Yu-Te Liao, Dan Neal, Josephine Esquivel-Upshaw, Stephen Pearton*

Breast cancer is the most commonly diagnosed cancer among women. Salivary biomarkers offer a non-invasive approach to detecting this cancer. In this study, a virus detection device based on metal-oxide-semiconductor field-effect transistors (MOSFETs) and functionalized, disposable test strips were utilized. The results demonstrated impressive limits of detection (LOD) and sensitivity when analyzing common breast cancer biomarkers, including HER2 and CA15-3. The LOD for these two biomarkers reached as low as $10^{-15}$ g/mL, and the detection of HER2 exhibited a high sensitivity of approximately 70/dec, enabling differentiation between HER2-negative cases. The sensitivity of CA15-3 was 30/dec, which was also able to distinguish between healthy individuals and breast cancer patients. Furthermore, the test takes less than a minute to determine the actual concentration of biomarkers in saliva. The technique is easy to operate and has the potential for widespread public use in the future.

## 213i Atomically Dispersed Supported Palladium Catalysts for High Performance Chemiresistive Gas Sensors

*Yaprak Ozbakir, Carlo Carraro, Roya Maboudian*

Incorporation of noble metals into chemiresistive gas sensors has been proven to be an effective strategy to enhance their sensitivity and selectivity for their use in air-quality monitoring systems. However, noble metal particles are prone to poisoning, resulting in catalyst deactivation. Furthermore, many such sensors require high operation temperatures to achieve the needed sensitivity. To overcome these issues, a strategy using atomically dispersed supported catalysts is proposed. Atomically dispersed supported catalysts (ADC) constitute a new class of materials offering the advantage of the most efficient use of metals combined with several unique properties, considerably different from their conventional nanoparticle equivalents. In this presentation, we report on the synthesis and characterization of Pd ADC supported on two support materials, namely nitrogen-doped graphene oxide (N-GO) and tin (IV) oxide ($SnO_2$). Our results indicate that the chemiresistive sensors fabricated using these Pd ADCs exhibit superior performance over their nanoparticle-based counterparts. Finally, we will correlate their structural and spectroscopic characterization with the observed catalytic efficiencies, and propose a response mechanism for their sensing behavior.

## 153b The Stability of ALD Thin-Film Coated Electrode Surface in Electrolyte Solutions

*Dorian Thompson, Yu Lei, Necmi Biyikli, Ilhom Saidjafarzoda*

Atomic Layer Deposition (ALD) thin films like $Al_2O_3$ and ZnO are extensively used in device fabrication. Assessing their stability in electrolyte solutions is vital to comprehend their performance and longevity in applications such as sensors, batteries, and fuel cells. Understanding film properties ensures device efficiency and durability in various electrochemical applications. While bulk metal oxides are stable in neutral pH electrolytes, this may not hold true for ultra-thin ALD films. In this study, we deposit ALD films of 1-10 nm thickness on electrodes and systematically examine their stability in representative electrolyte solutions. Electrochemical analyses will be supported by advanced characterization methods such as FTIR and AFM. This research will provide valuable insights into ALD thin film device stability in potential sensing and biosensing applications.

## 153c Design and Characterization of a 3D-Printed Vertically Stacked Sensor Array for Monitoring Electrochemical Signatures of Microbial Activity in Wetland Sediment

*Bukola Adesanmi, Sai Prasanna Chinthala, Zachary Cheney, Chelsea Monty-Bromer, John Senko, Lauren Kinsman-Costello*

Electron transfer processes in wetland sediment are driven by the activities of microbial communities and the surrounding environment, such as the reduction and oxidation of specific chemical compounds. To monitor the electrochemical signature arising from microbial activities, we have utilized a vertically stacked zero resistance ammetry (ZRA) sensor. ZRA is an electrochemical technique to measure electron flow between two electrodes immersed in electrolyte solution. The Vertically stacked ZRA is comprised of an array of identical graphite electrodes which are electrically connected through zero resistance ammeter. We have designed and produced the graphite electrode through additive manufacturing. We characterized the electrode by cyclic voltammetry, scanning electron microscopy (SEM) and X-ray diffraction analysis (XRD). We deployed the ZRA sensor to OWC NERR throughout the spring/summer of 2022 and 2023 and measured the current between graphite electrodes due to microbial activities in wetland sediment. We validated our result with dissolved oxygen data, surface water fluxes of $CO_2$ and $CH_4$, pore water and sediment analysis, and microbial community analysis. Our results indicate that the ZRA approach can be used to monitor electrochemical signature arising from microbial activities in wetland

## 153d Detection of Pyocyanin and Methylene Blue Using Electrochemical Impedance Spectroscopy (EIS) with Electrodeposited Nickel-Iron-Carbon Composite Electrodes

*Arash Bahrololoomi, Elizabeth Podlaha-Murphy*

The detection of pyocyanin (PYO), a metabolite of *Pseudomonas aeruginosa*, a gram-negative bacterium, is of interest due to its association with disease responsible for chronic infection in people with cystic fibrosis and in burn patients. Methylene blue (MB) can serve as a comparable model for PYO due to similar structure and redox potential. Methylene blue is a dye and because it is a pollutant its detection is also of interest in its own right. Both PYO and MB can be detected at redox potentials under diffusion control by conventional voltammetry, where the generated current is directly proportional to concentration. In contrast to reactions controlled by diffusion, the capacitance can also be used at the electrode-electrolyte interface to avoid the diffusion limit at low concentration. Electrochemical impedance spectroscopy (EIS) can capture the different behavior. EIS was used in a three-electrode cell configuration in a phosphate buffer solution (PBS) containing various concentrations of MB and/or PYO. Electrodeposited nickel-iron nanoclusters onto a glassy carbon electrode was fabricated to enhance charge transfer while at the same time permitting adsorption of the organic molecules to a glassy carbon surface. The Ni–Fe nanoclusters were prepared from a sulfate-boric acid electrolyte by a pulse galvanostatic method. Linear EIS equivalent circuit models were used to characterize the signal over a wide frequency range at different steady state potentials. The models included constant phase elements (CPE) that were used to determine an average, effective capacitance of the interface, resistances to represent charge transfer, and a Warburg diffusion impedance. The capacitance component of the equivalent circuits resulted in a larger signal as the analyte concentration decreased, in contrast to conventional voltammetry techniques under a diffusion control, enabling a much lower detection limit, below 1 mM. The controlling phenomena that altered the electrode-electrolyte capacitance was attributed to adsorption.

## 153e Paper Microfluidic Based Wearable Patches for Biomarker Sensing in Sweat

*Sneha Mukherjee, Sabrina Pietrosemoli Salazar, Tamoghna Saha, Michael D. Dickey, Orlin D. Velev*

Our group has introduced a simple and efficient platform for sweat sampling and handling based on osmotic-capillary principles and paper microfluidics. This innovative platform can harvest sweat noninvasively, even without active perspiration, through a combination of osmosis, capillary wicking, and evaporation. The hydrogel disk containing concentrated solute interfaces with the skin to enable osmotic sweat withdrawal, which is then transported via a paper strip using capillary wicking and disposed of on an evaporation pad. We have applied these principles to develop a range of simple and cost-effective wearable skin patches for analyzing potassium, lactate, and cortisol levels, based on lateral flow assays (LFAs). Potassium is a crucial intracellular electrolyte, and small variations in its range can cause serious health issues such as cardiac arrest. Our LFA platform uses a silicone and paper microfluidic conduit attached to a commercial $K+$ strip, and a polyacrylamide hydrogel with a higher osmotic strength than sweat. In-vitro testing on gelatin-based model skin allowed us to calibrate the platform with model sweat, and human trials have shown that it can detect $K+$ levels from human skin with very low sweat volumes (~2-3 μL) under moderate exercise and rest conditions. We have also demonstrated that our patch can efficiently sample sweat lactate, an essential biomarker for monitoring anaerobic metabolism, oxidative stress levels, muscle health, and tissue hypoxia in individuals. On-skin testing of the platform on both resting and exercising human subjects confirms that the patch can extract sweat and analyze changing lactate levels, which increase with exercise and muscle activity. Additionally, we are working on interfacing the paper platform with electrochemical sweat lactate sensors, allowing for long-term continuous electronic readout of results from the wearable interface. We are also developing wearable skin LFAs for detection of cortisol as a key stress biomarker in individuals, soldiers, athletes, and emergency personnel. The non-invasive wearable patches that we have developed can enhance affordable advanced healthcare, by eliminating the need to visit medical centers for biomarker-based diagnostics. These simple, facile, and inexpensive patches could allow numerous opportunities for at-home or in-field POC diagnostics.

## 153f Development of Sol-Gels and Nanoparticle Enhanced Sol-Gels for pH Sensing Performance

*Weiguo Xie*

Sol–gel is a wet-chemical process that involves the formation of an inorganic colloidal solution (sol) as the precursor and the gelation of the sol to form a three-dimensional network structure (gel). Sol-gel technology has been applied for the development of optical sensors and biosensors, due to 1) its chemically, photochemically, thermally, and mechanically stable; 2) optically transparent; 3) being formed at low temperature and under mild chemical conditions, allowing the entrapment of molecules with thermal and chemical sensitivity.

The sol–gel process has three major steps: 1) the raw materials (reactants and nanoparticles) were dispersed in the solvent by an ultrasonic treatment, 2) Tetraethyl orthosilicate (TEOS) hydrolyzed to form active monomers, which polymerized to be a sol, and 3) eventually, the sol started to form a gel with a certain spatial structure by aging. Figure 1 shows the intensity spectra for sol-gel pH sensor with cresol red indicator dye.

The incorporation of $TiO_2$ nanoparticles enhances the hardness and elastic modulus of the gel by 0.18 GPa and 3.7 GPa, respectively, so that it provides a good and solid support for the immobilization of the indicator for a better sensitivity and stability in the sensing film. The $TiO_2$-sol-gel film can be used to detect pH in a broad range of 1-12 fast and reversibly. It has advantages over the traditional electrochemical analysis with no charge, strong anti-electromagnetic interference, no pollution, and can be employed to develop optic chemical sensor for remote measurement and multi-channel detection. It can be envisaged that the sol-gels and nanoparticle enhanced sol-gels have great potentials to be the solid base for next-gen sensors.

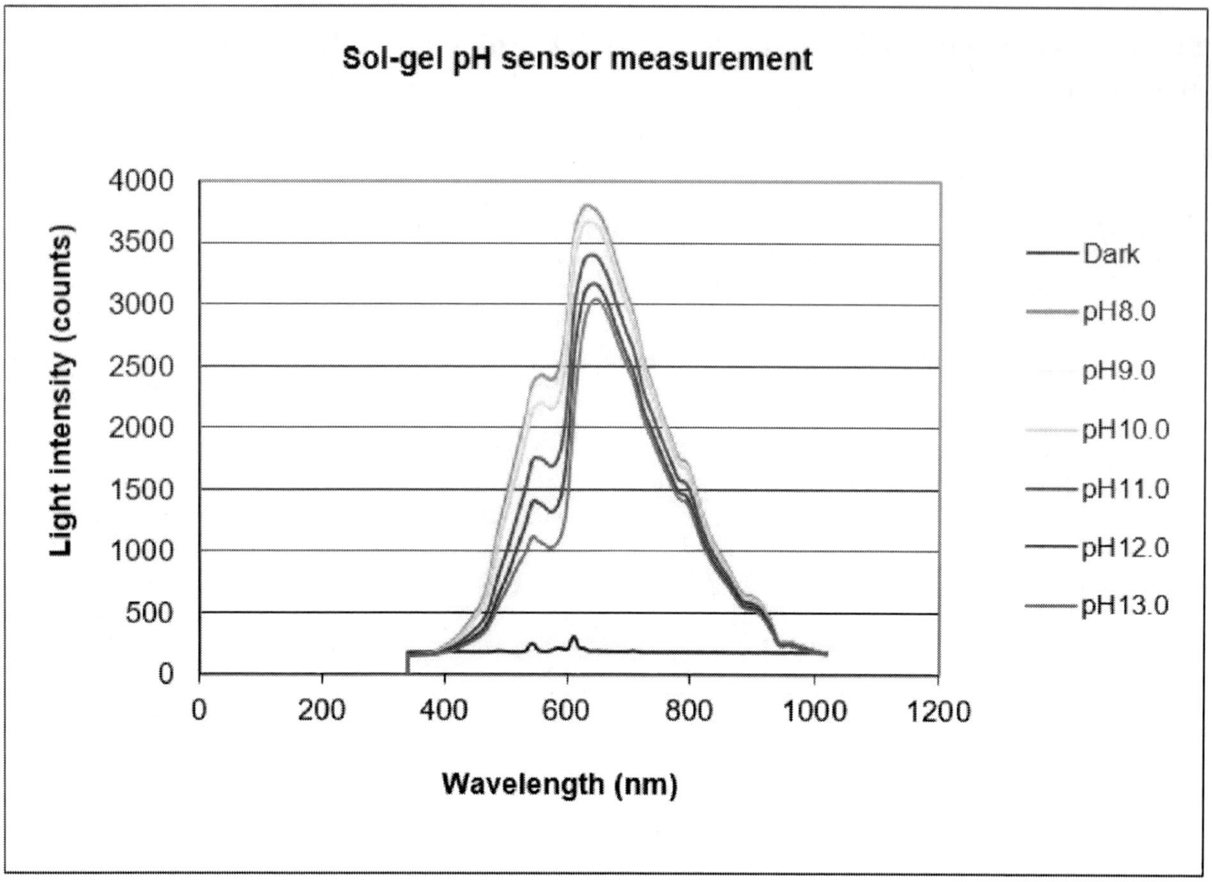


Figure 1: Intensity spectra for sol-gel pH sensor

## 269a Interpretation of Impedimetric Signals of Biosensors

*Jeffrey Halpern, Zahra Panahi*

Over the past several years, our group has been developing soft matter dynamic surfaces to act as chemical and biosensors. One surface is a cyclodextrin mediated surface that decouples when exposed to an analyte. Another surface uses elastin-like polymer temperature response to look at different surface states. Shared in common is that these surfaces and sensor construction is no electrochemical mediator exists to clearly interpret a faradaic signal commonly seen in voltammetry or amperometry.

My group has been actively working with electrochemical impedance spectroscopy (EIS) to interpret the different dynamic surface states without an electrochemical mediator. We have been relying on interpreting the capacitive and the out-of-phase oscillating current response in the EIS signal. Within this talk we will provide a fundamental review of EIS, how It can be used to look at different surface states, and the current interpretation of our surfaces.

Support was provided by a National Science Foundation EPSCoR award (#2119237).

## 269c Detection and Imaging of the Plant Pathogen Response By Near Infrared Fluorescent Polyphenol Sensors

*Robert Nißler, Sebastian Kruss*

Plants use secondary metabolites such as polyphenols for chemical defense against pathogens and herbivores. Despite their importance in plant pathogen interactions and tolerance to diseases, it remains challenging to detect polyphenols in complex plant tissues. Here, we create molecular sensors for plant polyphenol imaging that are based on near-infrared (NIR) fluorescent single wall carbon nanotubes (SWCNTs). We identified polyethylene glycol-phospholipids that render (6,5)-SWCNTs sensitive ($Kd$ = 90 nM) to plant polyphenols (tannins, flavonoids, ...), which red-shift (up to 20 nm) and quench their emission (~1000 nm). These sensors report changes in total polyphenol level after herbivore or pathogen challenge in crop plant systems (Soybean *Glycine max*) and leaf tissue extracts (Tococa spp). We furthermore demonstrate remote chemical imaging of pathogen-induced polyphenol release from roots of soybean seedlings over the time course of 24h. This approach allows in situ visualization and understanding of the chemical plant defense in real time and paves the way for sustainable plant phenotyping for optimized polyphenol secretion.

## 269d Development of a Gas Sensor Array for Detecting Toxic VOCs Using Metalated Gold Monolayer Protected Clusters

*Xiao-An Fu, Sujoy Halder, Prasadi Adhihetty, Michael H. Nantz*

Gas sensor arrays are a promising means of detecting toxic volatile organic compounds (VOCs) in environmental air. Among different airborne VOCs, benzene, toluene, ethylbenzene, xylene (BTEX), and trichloroethylene (TCE) are the most abundant and toxic VOCs. These VOCs induce cardiovascular metabolic syndrome and lung cancer, etc. Detection of these VOCs in environmental air by current gas sensors is very challenging because of sensitivity, selectivity and air matrix interference issues. Thiol-functionalized gold nanoparticle-based chemiresistors provide an advantage over other gas sensors, such as lower fabrication cost, room-temperature operation, and higher sensitivity and selectivity. This is achieved by tailoring the composition of the thiolate monolayers. The present work demonstrates development of alkali metal ion-carboxylate functionalized gold monolayer protected clusters (Au MPCs) as a sensing material for highly sensitive and selective detection of BTEX and TCE.

A gas sensor array has been fabricated using micro-electromechanical system (MEMS) technology. The sensor array features four circularly shaped sensing areas containing interdigitated electrodes that can perform simultaneously testing four different sensing materials. Lithium, sodium, potassium, and cesium carboxylate-linked Au MPCs were synthesized and tested for sensing vapors of BTEX, TCE as well as some non-aromatic and chlorinated VOCs. The results demonstrate that the responses of metalated Au MPCs are rapid, reversible, and linear for all the tested analytes. Cation-linked Au MPCs improve sensitivity and selectivity toward sensing BTEX due to cation–π interactions and the relative strength of the sensitivity follows the order of $K^+ > Na^+ > Li^+$. Furthermore, $Cs^+$-functionalized Au MPCs respond highly to chlorinated compounds, which helps to detect TCE in air. These metalated Au MPCs sensors show much promise for developing a sensor array for real-time measurements of toxic airborne VOCs.

# 269f Electro-Spun Core-Sheath Fibers with Polymer Network Liquid Crystal As Wearable Sensors

*Kun-Lin Yang, Zhibo Zhang*

Electro-spun polymer-liquid crystal (PLC) fibers have potential applications such as wearable sensors and adaptive textiles because of their rapid response and high flexibility. However, existing PLC fibers only have a narrow responsive range and poor resistance to heat and chemicals. Herein, a new type of PLC fiber is prepared by using a coaxial electro-spinning process. The core solution is 4'-pentyl-4-biphenylcarbonitrile (5CB), and the sheath solution is a mixture containing 13 wt. % PVP and 10 wt. % reactive mesogen (RM). After UV exposure of the fibers, 5CB in the core and RM diffusing from the core are cross-linked into an LC polymer. The fibers have a highly uniform morphology with an average diameter of 3.2 ± 0.5 μm, and mesogens inside the fibers align unidirectionally with the long axis of the fibers. The fibers show a broad phase-transition temperature range between 13.5 °C and 155.5 °C and have a response time of less than 10 s. The temperature range can also be controlled by adjusting components in the electro-spun fibers and UV exposure time. The core-sheath fibers prepared in such a manner exhibit excellent heat and chemical resistance with reversible optical responses. Moreover, when the fibers are exposed to volatile organic compounds (VOCs) such as toluene, the fibers show a rapid optical response to toluene vapor within 25 s. This study demonstrates that the fibers are potentially useful for preparing flexible temperature and chemical sensors.

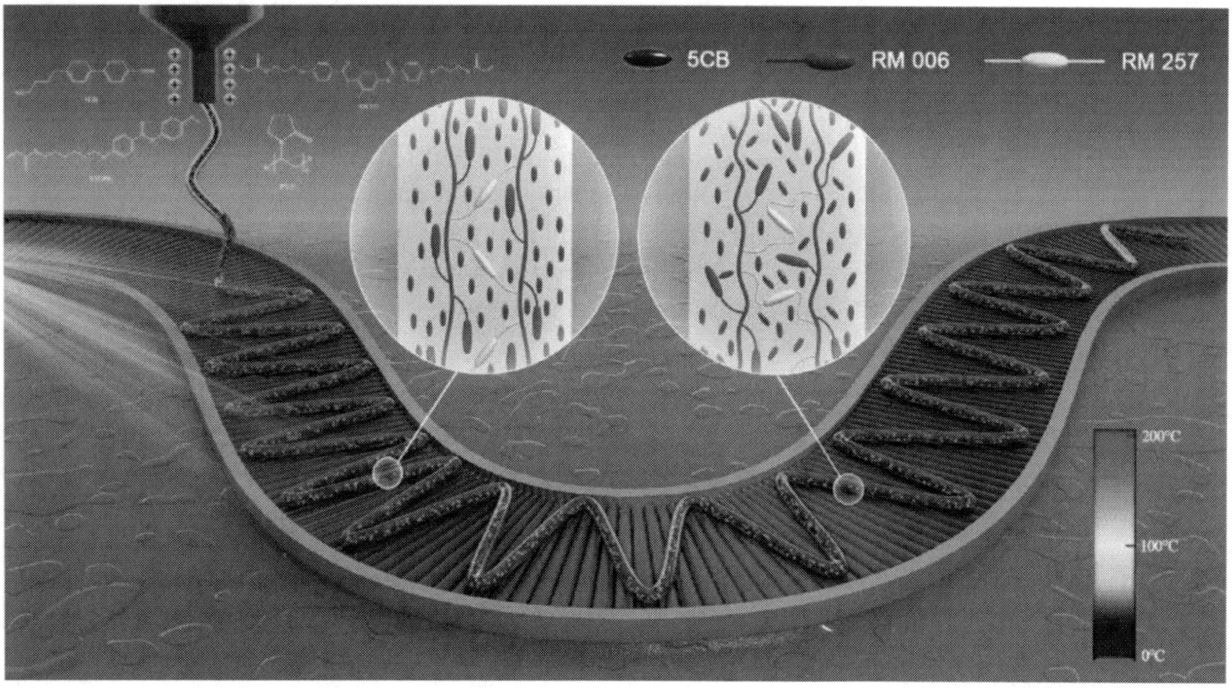

# 269g Application of Magnetic Phase Transition in Energy Harvesting Wireless Temperature Sensors

*Hikaru Kiyomoto, Yuka Sakai, Yasuki Kansha*

In recent years, many industries are showing great interest in sustainable societies, with concepts such as cyber-physical systems (CPS). In doing so, energy efficiency becomes a crucial point of concern. From the initial data acquisition to the data storage and communication component of a CPS, maintaining a sensible energy efficiency is of paramount importance for a sustainable society. Energy harvesting, where electricity is made from low-quality energy sources, is promising in this regard. Yet a common difficulty with many energy harvesters is that the energy output simply does not meet the energy demand in the system. With that said, it is nonetheless promising to be able to harvest the energy necessary for operation in the components of a CPS. Thus, an investigation in the energy efficiency of energy harvesters for sensor devices is a key topic in order to realise a sustainable society.

In this research, the authors proposed a novel energy harvesting wireless sensor with possible application for a temperature data acquisition device. This can for instance be for monitoring body temperature, which is not only quite important for various key components of our society, such as public health, but also is a highly quantifiable measure for monitoring health. The proposed device utilises the magnetic properties of materials as the working principle in order to produce electricity from low quality subject heat while simultaneously monitoring the temperature. When a magnetic material approaches its Curie Temperature, a magnetic phase change occurs, causing a corresponding magnetic flux change. By incorporating electromagnetic induction with solenoids, this magnetic flux change could be measured as voltage. Thus, this phenomenon can be applied for an energy harvesting temperature sensor in the near range of a material's Curie Temperature, as the voltage reading is representative of the magnetic flux change and thus the material temperature. Thus, with an apt material, the device has the potential to operate as a temperature sensor. In this research, the authors analyse the applicability of the proposed device for a large-scale temperature sensor, in terms of its quality as a temperature sensor and its energy harvesting capabilities. Namely, an investigation on the fundamental magnetic properties of the sensor probe material in question and the electromagnetic induction capabilities are conducted.

## 269h A Bayesian Approach to Curate the Optimal Sensing Elements for a Quantitative Gas Sensor Array

*Paul Morris, Gbenga Fabusola, Adam Lambert, Cory Simon*

Our objective is to showcase a Bayesian computational method to curate the optimal combination of sensing elements for a gas sensor array. We achieve this objective by treating the gas sensing task as a statistical inverse problem. As an illustrative example, we demonstrate our method by curating the optimal pair of zeolitic imidazolate frameworks for a gravimetric sensor array for $CO_2/C_2H_4$ sensing.

## 315a Outperforming State-of-the-Art Risk Factors of Cancer and Cardiovascular Diseases through Non-Invasive Liquid Biopsy Sensors Measuring Lipid-Protein Assemblies

*Sonu Kumar, Nalin Maniya, Satyajyoti Senapati, Hsueh-Chia Chang*

Cancer and cardiovascular diseases account for roughly 50% of all global deaths and can often be prevented, cured, or managed with early detection and treatment. The challenge associated with early detection lies in the invasiveness of the detection method, the highly sophisticated required machinery and overall, the high cost of the test. Still, even with the state-of-the-art facility, we cannot detect cancer in its initial stages since most of them rely on looking at methods such as tissue staining, performing biopsies and RNA sequencing, mammography, colonoscopy or the highly inconclusive MRI of the tumor cells – all requiring information on the location of the tumor which is unknown for the undiagnosed patients. This also has long-term side effects – drilling into a healthy patient's brain or spinal cord to check for glioblastoma will lead to a long-term damage. Similar complexity goes for cardiovascular diseases where plaques are difficult to see in their early stages, while cholesterol tests fail to accurately predict cardiovascular risk (~65% true positive and true negative rate). Lipid-protein-nucleic acid assemblies (exosomes, supermeres, lipoprotein, etc.) in the blood can allow us to overcome these issues since their cargo (proteins, miRNA, mRNA, etc.) share the signature of their parent cell and provides us with a window to non-invasively look at tissues from different parts of the body using a routine venous blood draw. For example, exosomes rich with neuronal proteins such as Enolase 2 (ENO2) and L1CAM can allow us to non-invasively investigate the brain for tumors or signs of neurodegeneration since these particles can cross the Blood Brain Barrier (BBB). Looking at proteins such as Carcinoembryonic Antigen (CEA) on these nanovesicles allows us to look at gastrointestinal health and thus colorectal cancers. Additionally, PON1 on HDL is hypothesized to be responsible for significant atheroprotective properties of HDL and may be a much better cardiovascular risk marker.

The bottlenecks associated with the measurement of these particles rely on several factors. Firstly, the gold standard for protein measurement, Enzymatic Immunoassays (ELISA), are incompatible with lipids due to redox interference from lipid peroxides on these particles. At the same time, flow cytometry-based methods cannot be extended to such small vesicles. Secondly, ELISA and other commonly used methods lack the sensitivity (>pM) to assess these particles accurately (~fM-pM concentration), requiring more sensitive methods. Thirdly, there can be significant competition from the soluble versions of these proteins or non-targets, which requires us to have a large dynamic range to operate away from sensor saturation. Lastly, the detection method should be fast (<1 hour) and cheap (~$1) so that everyone can afford these tests – perhaps even as an annual or semi-annual test. This would ensure we can detect the disease early, start the treatment early (>85 percent survival for stage I cancer detection) and use these markers to monitor the efficacy of treatment, remission, and recurrence. We have developed Charge-gated Anion Exchange Membrane (AEM)-based Electrokinetic sensors which is very sensitive (~fM), cheap (~$1), fast (<1 hour), and has a very large dynamic range and is less susceptible to environmental factors.

The AEM possesses an ion-selective property that permits only counter ions (anions) to pass through. When an electric field is applied, this generates a unidirectional ion flux, leading to ion depletion on one side of the membrane and ion enrichment on the other. As the voltage increases, a negative charge surface polarization occurs, which triggers an instability in the interfacial electroconvective vortex. The ion depletion and mixing of the vortex cause changes in ion current conductance, leading to the creation of a current-voltage curve (CVC) that displays distinct underlimiting, limiting, and overlimiting regions, each with varying differential conductance. Binding of negatively charged analytes to specific probes covalently linked to the depletion side of the AEM impedes the electroconvective instability as these charges are immobile. Consequently, the voltage responsible for the over-limiting current in the CVC shifts by several volts, while the underlimiting region remains unaffected. Moreover, the ion depletion and controlled washing facilitate the removal of assay inhibitors and control the ionic strength near the sensing surface, making the sensing signal independent of the ionic strength, pH, and chemical composition of the original sample. In general, the surface of the platform features an antibody that captures the desired particle. Charged particles can be detected without the need for labeling, resulting in superior sensitivity compared to Surface Plasmon Resonance. For neutral particles, a sandwich method is employed using negatively charged silica reporters. The platform has demonstrated the ability to quantify even a subset as small as 1/10000th of each particle category, owing to its broad dynamic range.

We have currently validated our strategy of detecting lipid-protein assemblies as well as our technology by assessing cardiovascular risk, detecting glioblastoma and colorectal cancers non-invasively with ongoing studies that include pancreatic, breast and lung cancers. We were able to achieve an impressive AUC ~ 0.99 with PON1-HDL in distinguishing Coronary Artery Disease patients from healthy controls and an AUC > 0.95 for cancer patients from healthy controls. We compared our platform to the state-of-the-art methods used by healthcare providers for the cardiovascular risk assessment, outperforming Cholesterol tests, Apolipoprotein levels, HDL-P levels (AUC~0.65) that accurately predicted CAD in about 60-70% of the patients compared to our platform that predicted >95% of the CAD patients and controls accurately.

**Publications:**

1. Kumar, S.; Maniya, N.; Wang, C.; Senapati, S.; Chang, H.-C., Quantifying PON1 on HDL with nanoparticle-gated electrokinetic membrane sensor for accurate cardiovascular risk assessment. *Nature Communications* **2023,** *14* (1), 557.

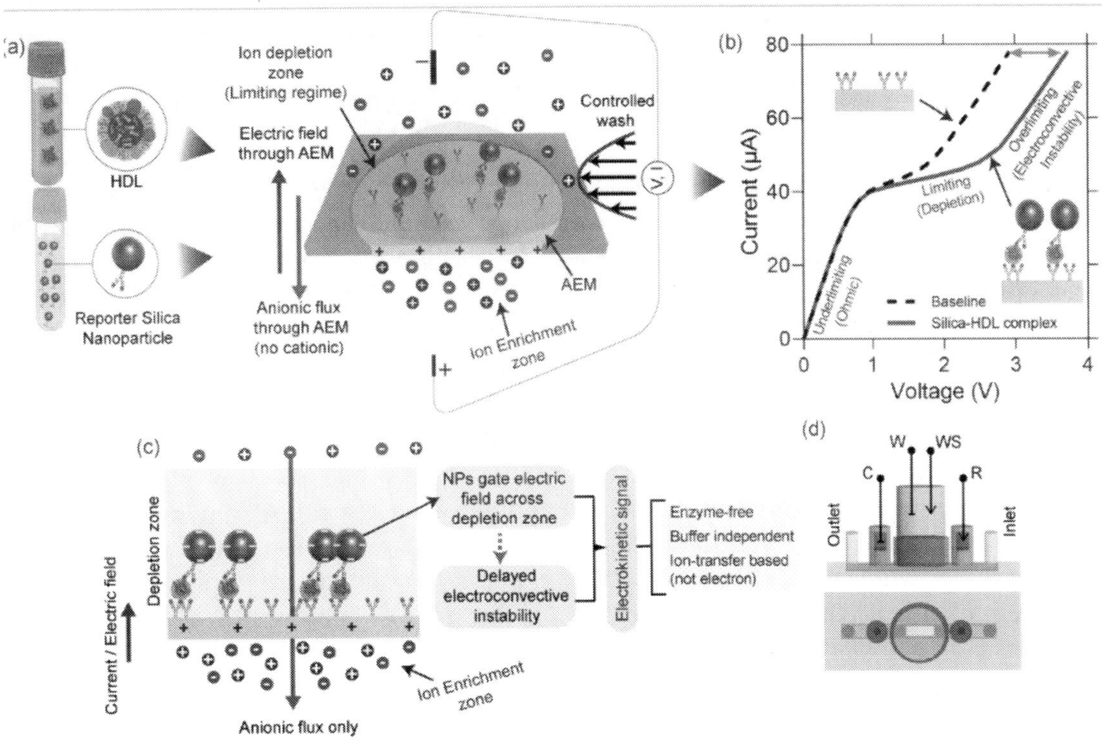
(a)
HDL
Reporter Silica Nanoparticle
Ion depletion zone (Limiting regime)
Electric field through AEM
Anionic flux through AEM (no cationic)
Controlled wash
V, I
AEM
Ion Enrichment zone
(b)
Current (μA)
Voltage (V)
Underlimiting (Ohmic)
Limiting (Depletion)
Overlimiting (Electroconvective Instability)
Baseline
Silica-HDL complex
(c)
Current / Electric field
Depletion zone
NPs gate electric field across depletion zone
Delayed electroconvective instability
Electrokinetic signal
Enzyme-free
Buffer independent
Ion-transfer based (not electron)
Ion Enrichment zone
Anionic flux only
(d)
W
WS
C
R
Outlet
Inlet

## 315b Rapid Sensing of Heat-Stress Markers in Organisms: CNN Based Learning Using Micrographs of Chicken Red Blood Cells Dispersed in Liquid Crystals

*Prateek Verma, Elizabeth Adeogun, Elizabeth S Greene, Sami Dridi, Ukash Nakarmi, Karthik Nayani*

An imbalance between bodily heat production and heat loss leads to heat stress in organisms. Heat stress is detrimental to all agricultural systems but is a particularly important stressor for the poultry industry; poultry entails fast growth and high yield, resulting in greater metabolic activity and higher body heat production, along with diminished well-being of the animal. When stressed, cells overexpress heat shock proteins (such as HSP70, a well-established intracellular stress indicator) and may undergo changes in their mechanical properties. Liquid crystals (LCs) respond to, and can be used to rapidly characterize, changes in mechanical properties of red blood cells; this was the driving principle for this research. Expression of HSP70 yielded a different LC response pattern as well as different cellular straining when compared to cells not expressing HSP70 when seen under an optical microscope. For rapid detection of such distinct cellular straining and LC patterns - scenarios where human judgement could be prohibitively difficult or slow - convolution neural network (CNN) based machine-learning (ML) models were trained on hundreds of such micrographs. Trained models exhibited remarkable accuracy of up to 99% on unseen microscope samples. Additionally, mechanical properties of both chicken and human red blood cells were altered by crosslinking them with glutaraldehyde and their micrographs were used to successfully test the CNN based classification methodology prior to the heat-stress experiments.

## 315c Aptasensors Based on Graphene Field-Effect Transistors for Arsenite Detection

*Jingwei LI, Abhishek Tyagi, Ting Huang, Hongwei Liu, Honglin Sun, Jiawen You, Md Masruck Alam, Xiangrong LI, Zhaoli Gao*

All-electronic chemical sensors based on field-effect transistors (FETs) hold tremendous promise for rapidly and sensitively detecting arsenic in groundwater. Their sensitivity, however, is fundamentally limited by a chemical gating mechanism where target ion charges are directly measured. We have developed a highly sensitive aptasensor, based on graphene FET arrays using CVD-grown graphene and the scalable photolithography process. The resulting aptasensor relies on the conformational change of the negatively charged aptamer upon arsenite binding, offering a wide analytical range, from 0.05 to 1000 ppb with excellent specificity against nonspecific ions, and also with a 0.02 ppb detection limit. Circular dichroism spectroscopy studies confirmed the conformational change of the aptamer, and molecular dynamic studies further validated that arsenite had bound with the hairpin loop of the aptamer, with higher interaction energy compared with other positions, signifying the location of the binding cavities of the arsenite ions and the aptamer. This work provides a highly accurate chemical sensing platform for the rapid detection of arsenite in drinking water.

## 315d Simplified Nano-Ities System for Simultaneous Detection of Metal Ions in Aqueous Samples

*Muzammil Nishar Ahmed, Anthi Savvidou, Pavithra Pathirathna*

Heavy metal toxicity is a growing concern worldwide as it causes fatal health disorders such as neurological issues and dysfunction of kidneys and livers. This problem is caused by rapid industrialization and poor recycling techniques of industrially produced metals that have left humans at high risk. To combat this problem, effective metal mitigation systems are needed. However, traditional techniques require expensive equipment and are difficult to operate, making them inaccessible to many. Moreover, the pre-treatment steps involved in these techniques limit real-time detection, and can alter the metal speciation, which is an important factor in determining metal toxicity. Thus, the development of a portable, low-cost, and reliable sensor that can provide accurate information on metal speciation is crucial.

The Pathirathna group had previously developed a Cd(II) sensor capable of detecting trace amounts of Cd(II) in environmental samples using ion transfer between two immiscible electrolyte solutions (ITIES). The sensor exhibited great stability and outstanding performance that was comparable to that of ICPMS. While detecting toxic metals in the environment, detecting multiple analytes would enhance the resourcefulness of the sensor. While extensive research has been performed using dual working electrodes to achieve simultaneous detection of multiple analytes, fabricating a double bore configuration is often challenging and can have potential interference and crosstalk. In this study we propose a single bore and single ionophore configuration with ITIES to achieve the same. This method has several advantages including simplicity in fabrication, less expensive and eliminates the need to integrate different ionophores.

We prepared borosilicate glass electrodes with an inner radius of 300 nm. We employed the differences in the half-wave potential, a signature electrochemical parameter for an analyte, to differentiate between multiple metals. We also demonstrated that the sensor can detect metals in artificial urine and artificial blood. To the best of our knowledge, this is the first time a single channel single ionophore based ITIES has been used for the simultaneous detection of multiple metal analytes. Our sensor shows great promise in being developed into an efficient metal monitoring device in the future. By offering real-time detection and accurate information on metal speciation, the sensor has the potential to revolutionize the field of heavy metal detection and help mitigate the rising global concern of heavy metal toxicity. Future research will be focused on automating the detection process in which the device will be able to complete the detection process without the need to be monitored. This novel approach has the potential to revolutionize the field of heavy metal detection and help mitigate the rising global concern of heavy metal toxicity.

# 315e A Stretchable Wireless Smart Bandage for Multiplexed and Multimodal Monitoring and Combination Treatment of Infected Chronic Wounds

*Ehsan Shirzaei Sani, Wei Gao*

## A Stretchable Wireless Smart Bandage for Multiplexed and Multimodal Monitoring and Combination Treatment of Infected Chronic Wounds

**Ehsan Shirzaei Sani[1], Wei Gao[1*]**

[1]Andrew and Peggy Cherng Department of Medical Engineering, Division of Engineering and Applied Science, California Institute of Technology, Pasadena, CA, 91125 USA

### Introduction

Chronic non-healing wounds are a significant burden on the healthcare system, affecting over 6.7 million people in the United States alone and costing over $25 billion per year. Current therapies often require surgical intervention, and overuse of antibiotics can lead to drug resistance. Here, we introduce a fully-integrated wireless wearable bioelectronic system that effectively monitors the physiological conditions of the wound bed via multiplexed and multimodal wound biomarker analysis, and performs combination therapy through electro-responsive controlled drug delivery for anti-inflammatory antimicrobial treatment and exogenous electrical stimulation for tissue regeneration. The system was evaluated in vivo over prolonged periods of time in different rodent models with infected diabetic wounds, showing substantial accelerated chronic wound closure.

### Methods

A highly flexible smart bandage was designed and fabricated by using microfabrication techniques. The flexible sensor patch contains arrays of multiplex sensors for mapping different physiological conditions of the wound.

### Results and discussion

Here we present the development of a wireless wearable bioelectronic system with a multimodal biosensor array for multiplexed monitoring of wound exudate biomarkers, a stimulus-responsive drug-loaded electroactive hydrogel, and a pair of voltage-modulated electrodes for controlled drug release and electrical stimulation. The wearable patch is biocompatible, mechanically flexible, stretchable, and skin-conformal, and is capable of real-time selective monitoring of a panel of crucial wound biomarkers in multiple rodent models. The wearable patch is equipped with an on-demand electro-responsive drug release system, loaded with an antimicrobial and anti-inflammatory peptide, and can be modified with different electroactive hydrogels to deliver several other positively or negatively charged drugs and biomolecules. The integration of an electrical stimulation therapeutic module could facilitate cell motility and proliferation, and ECM deposition and remodeling in the process of wound regeneration resulting in rapid and effective cutaneous wound healing. Overall, the custom-engineered fully-integrated wearable patch could serve as a more effective, fully controllable, and easy-to-implement platform for personalized monitoring and treatment of chronic wounds with minimal side effects.

Reference:

1. Shirzaei Sani et al., Sci. Adv., 9, eadf7388, 2023.

## 315f Gold Leaf Electrodes for Affordable Biosensing in Resource-Limited Settings

*Marjon Zamani, Ariel Furst*

Affordable, point-of-care diagnostics (POC) are critical to achieve health equity in low-resource settings (LRS). A prevalent sensing method for these diagnostics is electrochemical readout. Gold is commonly used as a substrate for electrochemical biosensors due to strong coupling with thiolated biomolecules. However, state-of-the-art gold electrode fabrication techniques are too laborious and costly for deployment in LRS. To make gold electrode fabrication possible in resource-limited settings, we have developed gold leaf electrodes that are affordable and easily fabricated without specialized equipment. In addition to being almost an order of magnitude cheaper than their commercially available counterparts, they also outperform commercially available screen-printed gold electrodes because they lack dopants that can interfere with analyte detection. The electrodes are paired with a CRISPR-based assay to detect human papillomavirus (HPV) from clinical samples. They have also been used to distinguish between respiratory viruses. The versatility of our work illustrates that gold leaf electrodes are a platform technology that can be expanded to detect any disease in LRS.

## 315g New Electrochemical Sensing Platforms to Detect Ultra-Low Concentrations of Toxic Metals

*Pavithra Pathirathna, Noel Manring, Muzammil Nishar Ahmed*

Exposure to toxic metal ions is rising due to increased urbanization and industrialization; thus, humans are at a greater risk of getting metal toxicity due to exponent bioaccumulation along the food chains. Although the detrimental heal hazards associated with metal toxicity are well documented, there's a lack of proper analytical tools that can provide reliable qualitative and quantitative information about toxic metal ions due to several reasons, such as the complexity of metal chemistry and the limitations of fabrication and analysis protocols of sensors. Therefore, most existing methods rely on the measurements taken on postmortem tissues or urine and blood samples collected from the patients, thus, missing real-time information about the metal speciation and interactions. Moreover, a tool capable of early detection would be ideal for starting early therapeutics, thus preventing further accumulation and minimizing damage to vital body organs. In this vein, we develop different types of novel electrochemical sensors that are redox-based and non-redox based to detect ultra-low concentrations of toxic metals in complicated, aqueous buffers that mimic body fluids or in artificial blood and urine samples. Our nanoelectrodes based on ion transfer across two immiscible electrolyte solutions (ITIES), can detect Cd(II) ions with great sensitivity and excellent stability in complex matrices, including artificial urine and blood samples. We also engineered a novel, surface-modified carbon-fiber microelectrode (CFM) by electrodepositing dopamine onto the electrode surface to detect ultra-low concentrations of toxic metal ions in real-time using fast scan cyclic voltammetry (FSCV). Our surface-modified CFMs significantly improved the limit of detection and sensitivity of Cu(II) detection compared to bare CFMs. Moreover, we fabricated a double-bore CFM with two sensing components that can simultaneously measure two neurotransmitters and toxic metals via FSCV without altering the detection window of each analyte. To the best of our knowledge, this is the first electrochemical sensor to detect multiple analytes at ultra-fast scan rates. Our preliminary data obtained with these sensors showcases a great possibility of developing these sensors for future in vivo applications.

## 315h Temporally and Spatially Resolved Concentration Profiling Via Distributed Colloidal Particles with Onboard 1D Memristive Arrays

*Matthew Manion, Joshua Doctor, Albert Liu*

Accurate mapping of chemical concentrations in reactors is a problem as old as the discipline itself and remains an obstacle to establish complete systems-level insight and control. This issue extends beyond traditional reactor design to biological (*in vivo* and *in vitro)*, environmental, and other inaccessible systems of interest. Memristive materials are gaining traction as valuable tools in complex sensing applications, and encapsulating these materials into arrays on microparticles demonstrates the distinct potential for diffuse measurements of chemical, temporal, and physical data. In this study, we establish a workflow for the simulated deployment of these microparticle trackers in contemporary, sequestered systems of interest, measuring short chain fatty acid (SCFA) concentrations in the intestines, and how the simulation can aid in designing memristive particles for real-world use. We utilize, and compare, multiple Gillespie Kinetic Monte Carlo (GKMC) algorithms to allow for individual treatment of these particles, which reflects their distributed nature. Our results show that multiple copies of identical 1-D memristive arrays, when distributed inside a system of colloidal particles, can provide accurate SCFA concentration profiling in the intestines. This was previously believed to be the unique capability of the 2-D memristive array, which increases fabrication complexity and energy demands. Our results also show that by combining different types of memristive particles, one can extract more detailed information with fewer total particles deployed. This work builds upon previous ideas in memristive particle deployment and demonstrates complex information gathering capabilities with manufacturable 1-D arrays in isolated systems.

## 315i 30-Seconds Sars-COV-2 Human Sample Diagnosis and Analytical Specificity Analysis Using Disposable Strips on a Mosfet Platform

*Chao-Ching Chiang, Chan-Wen Chiu, Minghan Xian, Fan Ren, Cheng-Tse Tsai, Yu-Te Liao, Josephine F. Esquivel-Upshaw, Stephen Pearton*

A SARS-CoV-2 rapid detection technology employed commercially available test strips similar to the blood glucose test strips were developed with a detection time of 30 sec and the sensitivity as good as polymerase chain reaction (PCR). The analytical specificity of this technology was also investigated by testing the cross-reactivity and microbial interference on this biosensor platform using FDA suggested 31 different contaminants. Variants of SARS-CoV-2 have mutated along with the worldwide spread of the pandemic. Currently, the most common detection methods are PCR and lateral flow tests. The former takes more than an hour to obtain results and the latter has difficulty detecting the virus at low concentrations (for cycling threshold values, CT values, >25 to 27). Therefore, the demand for a fast, cost-effective, and low detection limit testing method has been significantly increased. In this work, 60 human saliva samples with 30 positive and 30 negative were collected and tested with PCR for SARS-CoV-2 prior to using our disposable strips. For this technology, the test strip was connected to the gate electrode of the MOSFET on the printed circuit board that serves to amplify signals. A synchronous double-pulsed bias voltages (around 1 ms) were sent to both the drain and gate of the MOSFET. The resulting change in drain waveforms was converted to digital readings and compared with cycling threshold (Ct) values of human samples to assess this sensor technology. The result signifies that positive human samples with a range of Ct values from 17.8 to 40 can be differentiated, along with proving none of those high-risk organisms would hinder the sensitivity of our system. Which demonstrated the potential of this system to be developed into a cost-effective and point-of-care rapid detection for SARS-CoV-2.

## 315j Accessing Sweat without Perspiration: The Journey Towards Exertion-Free Sweat Sampling, Management, and Sensing

*Tamoghna Saha, Michael Dickey, Joseph Wang, Orlin D. Velev*

Sweat is an essential biofluid for monitoring individuals' health as it contains several key biomarkers from blood. Additionally, sweat is relatively easy to generate on-skin due to the wide distribution of sweat glands throughout the body.[1] Conventional sweat based devices have commonly relied on active perspiration release - such as one caused by physical exertion, thermal stimulation, or iontophoresis, for accessing the biomarker information in sweat. These techniques make such devices susceptible to dilution and contamination from excess sweat amounts and be inoperative on sedentary subjects and under the conditions of low humidity. Hence, the focus of this talk will be on our two recently developed passive sweat accessing techniques that do not require external sweat presence on-skin prior testing – osmotic and diffusion-based sampling. The osmotic based sweat extraction technique utilizes a hydrogel of lower chemical potential (of water) than sweat for extraction. By combining this technique with evaporation assisted pumping, we have developed wearable patches to operate on skin for long-term (~ hours) sweat lactate monitoring.[2,3] The diffusion-based technique relies on the natural perspiration rate of the body, which is dominant in areas with high sweat gland density, such as the fingertip. A porous hydrogel can facilitate biomarker transport from the fingertip to the transducing element, eventually allowing discrete, touch-based, and painless monitoring. For this talk we will mainly discuss how we have utilized this technique for ketone detection in sweat due to its correlation to diabetes.[4] Overall, the successful transition of these techniques to point-of-care and wearable devices holds great potential in laying the foundation for next generation sweat based devices for personalized healthcare monitoring.

References:

[1] T. Saha, R. Del Caño, E. la De Paz, S. S. Sandhu, J. Wang, *Small* **2022**, *2206064*, 2206064.

[2] T. Saha, J. Fang, S. Mukherjee, M. D. Dickey, O. D. Velev, *ACS Appl. Mater. Interfaces* **2021**, *13*, 8071.

[3] T. Saha, T. Songkakul, C. T. Knisely, M. A. Yokus, M. A. Daniele, M. D. Dickey, A. Bozkurt, O. D. Velev, *ACS Sensors* **2022**, *7*, 2037.

[4] J. Moon, R. Del Caño, C. Moonla, K. Sakdaphetsiri, T. Saha, L. Francine Mendes, L. Yin, A. Chang, S. Seker, J. Wang, *ACS Sensors* **2022**, *7*, 3973.

## 365a Exploring Film-Coated Electrodes for Pfas Detection

*Dorian Thompson, Yu Lei, Heejeong Ryu*

Per- and polyfluoroalkyl substances (PFAS) are an emerging group of anthropogenic contaminants that are arduous to degrade and possess a horde of detrimental health effects in mammals. Consequentially, exposure health advisory limits of these substances are extremely low, such as the US Environmental Protection Agency advising water levels of perfluorooctanoic acid (PFOA) and perfluorooctane sulfonate (PFOS) to be below 0.004 parts per trillion (ppt) and 0.02 ppt, respectively. To easily sustain healthy levels of these chemicals, PFAS detection platforms must have ultra-sensitivity, along with low cost, portability, and a fast response time. In this study, we functionalize electrode with nonconductive film and investigate its suitability for PFAS sensing. Utilizing molecular printing technologies, the films create cavities that expose the electrode's active surface and attract PFAS molecules. Since PFASs lack electroactivity, successful PFAS capture alters the collected electrochemical signal. The optimization of parameters such as film composition, structural stability, and sensing sensitivity will be explored, and the sensor's performance will be presented and discussed.

## 365b A Wearable Fabric Sensor for Monitoring Hydration in High-Level Athletes

*Victoria Stege, Zachary Cheney, Rebecca Ellis, Michael Fulmer, Chelsea Monty-Bromer, Mary Pat Nicodemus, Ronald Otterstetter*

Currently, there is a need for more accurate and more convenient methods of sweat sodium testing for endurance athletes. Sweat sodium is known to have a notable correlation to body fluid balance, making it an ideal way to detect and monitor overall hydration status. However, traditional sweat sodium monitoring can be inaccurate as it requires sample collection and analysis to be performed in separate steps. This research focuses on a newly designed wearable fabric sensor (WFS). The sensor consists of a non-woven fabric base that is treated with dispersed carbon nanotubes. The carbon nanotubes on the fabric are then functionalized with a size selective molecule for sodium detection. After calibration and assembly, the WFS can be worn by the athlete during exercise. As sodium binds to the WFS, the change in resistance is measured and output via custom-made Bluetooth electronics. After the test, the athlete's steady-state sodium loss, time to start sweating, and time to reach steady-state sodium loss are all determined. Because of its unique design, this WFS can jointly collect and analyze sweat samples during exercise and can improve the ease and accuracy of on body sweat analysis. During this study, subjects (n = 31, ages 18-60 years) wore a WFS during a 30-minute cycle ergometry test at about 65% heart rate reserve or 90-minute running test at moderate intensity. The participant's sweat was collected every 15 minutes on the lower forearm using sweat collectors. Additionally, subjects wore the wearable fabric sensor (WFS) on the upper forearm throughout the duration of the test. The sweat collected was analyzed using a commercial sweat sodium analyzer (SSA). Comparisons were made between the measurements from the SSA and from the WFS. Accuracy was determined by comparing predicted WFS concentration to the actual concentration from the commercial SSA and analyzed statistically using ANOVA and Bland-Altman plots. Bland-Altman analysis of the cycling data showed the average concentration difference between the WFS and the SSA was 5.35 mM, with 99% of data points between +/- 1.96 times the standard deviation (SD). Additionally, Bland-Altman analysis of the running data showed an average concentration difference between the WFS and SSA of 0.83 mM, with 92% of data points between +/- 1.96 times the SD. For both the running and cycling trials, the WFS showed similar average sweat sodium values as compared to the SSA. These analyses show that there is no statistical significance between the values reported by the WFS and the SSA. In summary, the WFS accurately predicted sweat sodium concentration during endurance exercise when compared to an SSA.

This work was funded by NSF [grant number 1843539], Ohio Third Frontier TVSF [grant number 20-0004], and NSF [grant number 2111983]

## 365c Near-Infrared Fluorescent Carbon Nanotube Sensors for the Plant Hormone Family Gibberellins

*Kulaporn Boonyaves, Mervin Ang, Minkyung Park, Jianqiao Leslie Cui, Duc-Thinh Khong, Volodymyr Koman, Xun Gong, Thomas K. Porter, Seo Woo Choi, Kwanghun Chung, Nam-Hai Chua, Daisuke Urano, Michael Strano*

Gibberellins (GAs) are a class of phytohormones, important for plant growth, and very difficult to distinguish because of their similarity in chemical structures. Herein, we develop the first nanosensors for GAs by designing and engineering polymer-wrapped single-walled carbon nanotubes (SWNTs) with unique corona phases that selectively bind to bioactive GAs, $GA_3$ and $GA_4$, triggering near-infrared (NIR) fluorescence intensity changes. Using a new coupled Raman/NIR fluorimeter that enables self-referencing of nanosensor NIR fluorescence with its Raman G-band, we demonstrated detection of cellular GA in Arabidopsis, lettuce, and basil roots. The nanosensors reported increased endogenous GA levels in transgenic Arabidopsis mutants that overexpress GA and in emerging lateral roots. Our approach allows rapid spatiotemporal detection of GA across species. The reversible sensor captured the decreasing GA levels in salt-treated lettuce roots, which correlated remarkably with fresh weight changes. This work demonstrates the potential for nanosensors to solve longstanding problems in plant biotechnology.

# 365d Deciphering the Correlation between Color Response, Temperature, and Relative Humidity in a Photo-Patternable Polymeric Nanofilm for Temperature and Humidity Sensing

*Md Nayeem Hasan Kashem, Wei Li*

Relative humidity (RH) and temperature responsive nanofilms have been developed as colorimetric sensors based on light interference, with the sensors providing rapid and easy detection, high accuracy, good mobility, wearability, and power-free sensing ability. However, current studies on RH and temperature-sensitive colorimetric films focus only on the effect of either temperature or RH on the sensing ability. With an inherent relationship between temperature and RH, this constraint leads to significant limitations for the use of thin film sensors in practical applications. In this work, we address these limitations by examining the correlation between structural color (thickness), temperature, and RH using nanolayered polyelectrolyte films made from chitosan (CHI) and carboxymethyl cellulose (CMC). We propose a novel 3-dimensional (3D) non-linear regression model to describe the correlation between the normalized thickness, temperature, and RH; with the thickness prediction showing high agreement with experimental results. Furthermore, photoreactive carboxymethyl cellulose-azido derivative (CMC-$N_3$) was used to build CHI/CMC-$N_3$ films that can covalently crosslink upon UV irradiation [1]. By taking advantage of different swelling behaviors of CHI/CMC-$N_3$ films before and after UV treatment, a variety of color and crosslinked patterns are hidden/displayed at selected RHs and temperatures. This enables a tunable multimodal display of the pattern-encoded information. Remarkably, in temperature ranges from 2°C to 55°C, uniform colors covering the entire visible spectra were observed, while in the low temperature (2-10°C) and high RH (~80%) regimes, the film demonstrated vivid color changes with 1°C resolution.

The project was further extended to monitor and record of subzero temperatures up to -30°C based on the color response. Specifically, the film was enclosed in a sealed petri dish to control the amount of moisture condensed on the film. Consequently, it displayed uniform colors at (0 to -30°C) temperatures. The color changing behavior was reversible in response to temperatures and RH. By introducing moisture absorber inside the petri dish at subzero temperatures, the color changing behavior became irreversible. After placing the film at room temperature for a certain period, the moisture was absorbed, hence, the film did not change the color in even after placing back at subzero temperatures. The moisture absorbing behavior was dependent on the absorber type, weight of absorber, storage temperature and RH inside the petri dish. Hence, a wide range of temperature history recording technique was demonstrated.

Due to the attractive properties of the nanofilm, we expect the film to provide a versatile platform for potential applications in environmental sensing, anticounterfeiting, food storage, and quality control in drug transport.

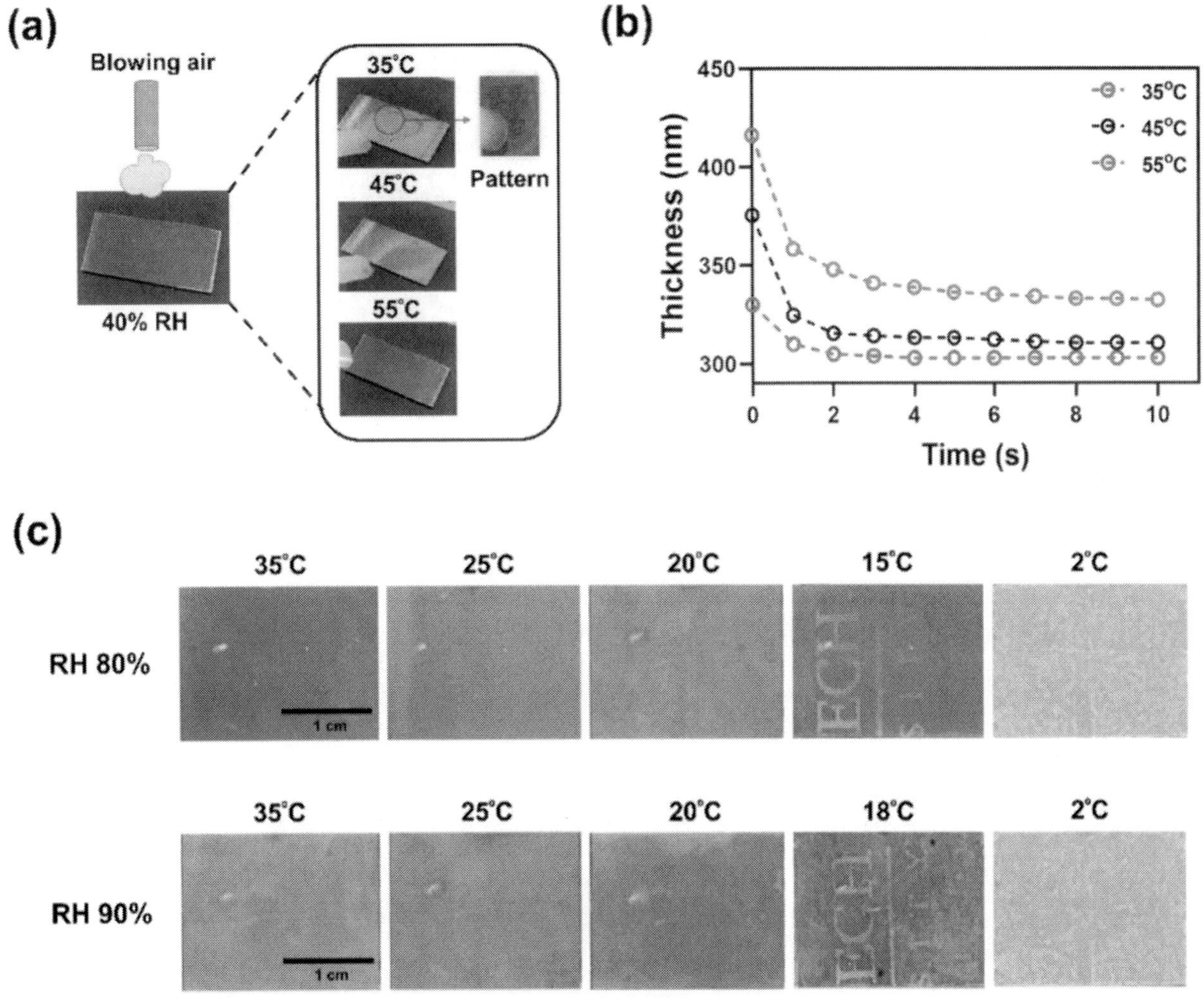
(a)
Blowing air
40% RH
35°C
Pattern
45°C
55°C
(b)
Thickness (nm)
Time (s)
35°C
45°C
55°C
450
400
350
300
0
2
4
6
8
10
(c)
35°C
25°C
20°C
15°C
2°C
RH 80%
1 cm
35°C
25°C
20°C
18°C
2°C
RH 90%
1 cm

## 365e Preferential *Trans*-Cleaving Behavior of CRISPR-Cas12a Towards 3'-Overhang dsDNA Unlocks to Develop Ultrasensitive Hybrid DNA Reporter

*Noor Mohammad, Megha Gongireddy, Qingshan Wei*

The CRISPR-Cas12a system is generally considered to have no *trans*-cleavage activities towards blunt-end dsDNA substrates. The recent discovery of indiscriminate *trans*-cleavage of dsDNA substrates (e.g., λ DNA) promotes more studies on its actual mechanism. Here, we observed that a short toehold at the end of the blunt dsDNA could behave as a molecular switch, activating the substrate's *trans*-cleaving activity. Additionally, we surprisingly discovered that the location of the overhang played a significant role: only a 3'-overhang could promote a complete *trans*-degradation of dsDNA substrate, while the 5'-overhang did not or only had a very mild effect. We hypothesized that the unique preferential *trans*-cleaving behavior of CRISPR-Cas12a towards 3' overhang dsDNA might be governed by RuvC and Nuc domains. Utilizing this newly found behavior of Cas12a, we designed a hybrid DNA reporter, which sensitively detected ssDNA targets with a LOD of ~250 fM. This study not only casts a new insight into the mechanism understanding of Cas12a biology but also provided a new class of hybrid dsDNA reporters for more efficient CRISPR-Dx.

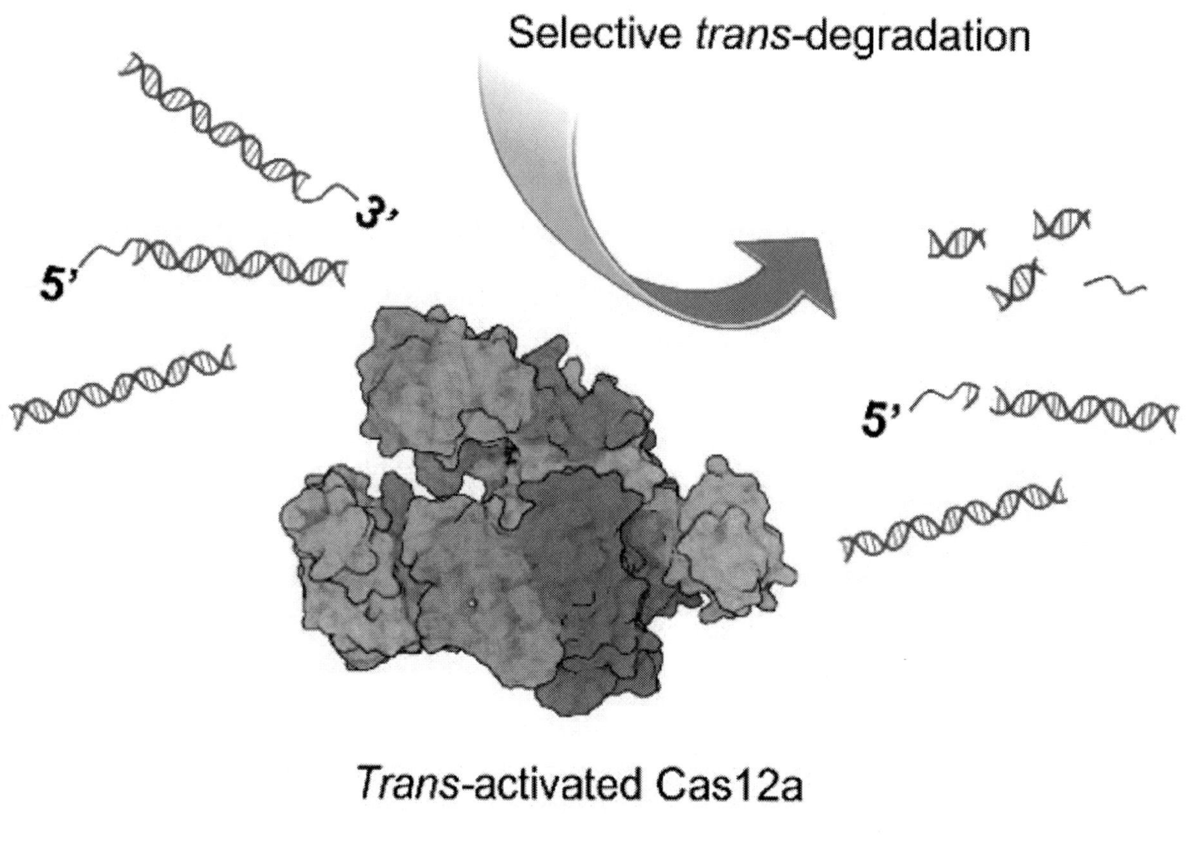


Figure 1: CRISPR-Cas12a-based directional *trans*-cleavage.

## 365f Double-Bore Carbon Fiber Microelectrodes for the Simultaneous Detection of Heavy Metals and Neurotransmitters Via Fast Scan Cyclic Voltammetry

*Noel Manring, Muzammil Nishar Ahmed, Pavithra Pathirathna*

The prevalence of neurodegenerative diseases (NDDs) is on the rise with more than 600 different conditions affecting over 50 million Americans each year. Due to the complex multifactorial etiology of these diseases, the development of therapeutics capable of slowing down or preventing further progression has proven challenging. Although the mechanism of these diseases is not well understood, the identification of abnormal levels of neurotransmitters has been linked to patients suffering with these disorders. These abnormal levels are correlated to the loss of motor functions, dementia, and overall cognitive incline that is associated with these diseases since neurotransmitters are responsible for neural communications. Additionally, the presence of redox active metal ions, such as copper, has been linked to increased neuronal cell death, a known hallmark of these diseases. Electrochemists have extensively used carbon-fiber microelectrodes (CFMs) to reveal significant findings related to NDDs over the last few decades. However, most studies analyze one neurotransmitter or metal at a time with a single CFM, thus, limiting the real-time information about interactions between neurotransmitters and other factors. In contrast, a multi-bore microelectrode with multiple carbon fibers (CF) can provide information about the heterogeneity of an evoked neurotransmitter and interactions between neurotransmitters at a specific location. However, engineering a microelectrode with multiple sensors has been challenging due to the lengthy processing time and cost associated with conventional fabrication and analysis methods. In this study, we fabricated a simple double-bore CFM with two sensing components. Each bore contains a single CF that acts as a single electrode capable of detecting a specific analyte using a unique waveform. We characterized our sensor with dopamine, serotonin, ascorbic acid, and Cu(II) with fast-scan cyclic voltammetry (FSCV). We acquired FSCV signals simultaneously in analyte mixtures prepared with two analytes in tris buffer. We constructed calibration curves showing that our double electrodes are not interfering with each other at nm-wide separation while cycling the potential at ultra-fast scan rates. We also found some exciting interactions between ascorbic acid and Cu(II). To the best of our knowledge, this is the first study reporting use of double-bore CFM to detect two analytes simultaneously without altering any electrochemical parameters at the temporal resolution of 100 ms; thus, showcasing an excellent possibility to use this sensor in vivo for real-time detection of multiple neurotransmitters and other toxic metal ions.

## 365g Electrodeposited Prussian Blue-Molecularly Imprinted Polymer Composite for Reagentless Lactate Detection

*Grace Dykstra, Yixin Liu*

The development of sensitive, selective, and stable sensor materials for specific biomarker detection aims to contribute to self-health monitoring and management. Molecularly Imprinted Polymers (MIPs) mimic natural systems such as antibodies with biomimetic recognition through polymerization in the presence of specific biomarkers. Their fabrication consists of polymerization and subsequent template removal which provides molecular cavities for detection. Electropolymerized MIPs (eMIPs) are polymerized via electrochemical methods and offer advantages of simple fabrication, quick detection and response time, and low-cost fabrication of biosensors at a reduced cost due to their inexpensive and quick fabrication methods and materials when compared to traditional antibody and enzyme based biosensors. In this work, eMIPs were prepared via electropolymerization of 3-aminophenylboronic acid (3-APBA) and pyrrole (py) with imprinted lactate molecules on the prussian-blue (PB) coated screen-printed carbon electrodes (SPCE). The electrode fabrication began with electrodeposition of prussian blue onto the surface as an embedded redox probe, followed by a MIP layer that both protects the PB nanoparticles and provides selective recognition for lactate molecules via the molecular imprinting sites. A major advantage of the PB nanoparticles deposited upon the surface is that it allows for reagentless analysis, where no external redox probe is required for electrochemical detection. Lactate levels were tested between 1nM and 1mM with high sensitivity and selectivity on the PB-based eMIP. Direct electrochemical detection of biomarkers through PB-based eMIPs can facilitate quicker diagnostics and enhance personal healthcare monitoring in the near future.

# AUTHOR INDEX